I0487875

Spinal
Stabilization

A Functional Rehab Program

Spinal Stabilization

A Functional Rehab Program

Douglas J. Taber, DC and
Douglas Van Vorst, DC

Copyright © 2010 Douglas J. Taber, DC and Douglas Van Vorst, DC

ISBN 978-0-557-15847-8

This booklet is to be used to reinforce appropriate spinal stabilizing exercises, assuming the patient has been evaluated by a licensed spinal healthcare provider and already shown these exercises. The authors recommend that patients seek a proper functional evaluation and subsequent instruction

from their doctor or therapist. This is not a patient-guided home exercise program or fitness regimen, and is not intended as a substitute for professional healthcare. Discontinue any exercise that causes numbness or tingling sensations or if your symptoms increase.

DISCLAIMER

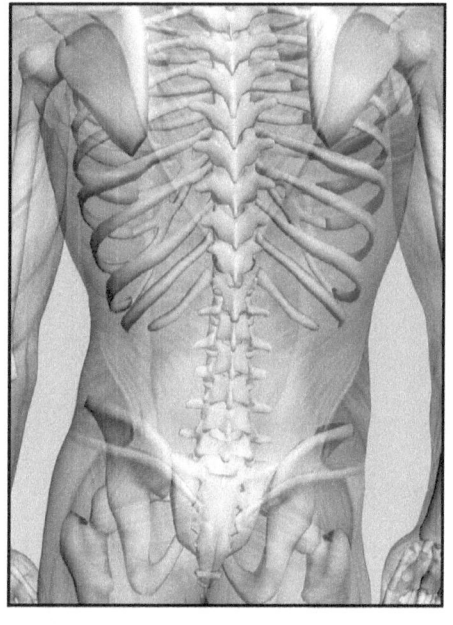

The information provided is to be used as an aid for clinician and patient. It is to be used at the discretion of the clinician and expresses no guarantee of results. This information is the copyrighted intellectual product of the authors. The information is the informed opinion of the authors and is backed by cited research, years of clinical experience, and actual patient encounters. The prescribing clinician is solely responsible as a licensed professional to interpret this information and apply your own judgment as to the validity and applicability of the information contained herein according to the Federal Law and state laws, including any constraints on your scope of practice. Every effort has been made to ensure the accuracy of the contained information. No patient should perform any of these exercises without proper instruction from their health care provider. The authors disclaim any responsibility for any adverse effects resulting directly or indirectly from suggested exercises, undetected errors, or the readers' misunderstanding of the text contained herein.

All rights reserved. Copying, disseminating, or distribution of this material in any form, including Photostat, microfilms, xerography, or any other means is strictly prohibited. This material may not be incorporated into any electronic or mechanical retrieval system without written permission of the authors and copyright owners.

TABLE OF CONTENTS

Spinal Stabilization: A Functional Rehab Program

INTRODUCTION

What are Spinal Stabilization Exercises?

Although conservative treatments are often useful in reducing pain and nerve compression, research indicates that delayed activation, or dysfunction in neurological control of the deep spinal and abdominal ("core") muscles often follows an episode of low back pain.[1-7]

In a healthy back, specific core muscles allow movement but also maintain alignment of the vertebrae. These muscles fire when movement is anticipated. This is known as **preactivation.** The key to optimal muscular spinal stability is the orchestration of activation, not strength, of these deep core muscles. This recruitment deficiency may not return to normal in the long term without targeted exercises like the ones in this booklet.

Like a boat mast, the deep spinal and core muscles provide stability to the entire spinal column, allowing orchestrated support and flexibility.

Essentially, these exercises are meant to retrain the speed at which your muscles react to forces that tend to injure an unprotected or weakened spine. Your doctor will guide you through this three phase program, progressing you from relief positions and core awareness principles into a home-based course of therapeutic exercises that will help you retrain weakened or de-conditioned muscles and re-establish normal motor control. This program is aimed at regaining normal global spinal function and minimizing risk of chronic or recurring low back pain.

Anatomy for the Patient: Basic BioMechanical Concepts

The spine is made up of multiple segments of vertebrae (bone), discs (spacers/ shock absorber analogous to jelly donuts), and soft tissue consisting of ligaments and muscles, which hold the back together and allow it to move. These structures all have nerve innervation (electrically wired) and are networked, designed to work together.

One of the most important structures is the disc. When you bend forward at the spine the disc is squeezed by the vertebrae above and below causing the jelly substance inside the disc to bulge backwards.

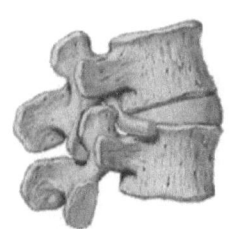
If this is done over and over and the spine is not held in a neutral position with supporting muscles of the abdomen and stomach activating in concert, the jelly inside the disc pushes backwards damaging the back of the disc and eventually this can tear the disc and herniated i.e. A herniated disc [1,2,3,4]

Discs hydrate (increased water pressure of the jelly) at night while you are sleeping. In other words there is an increase in disc pressure when you first wake up. Much like a bicycle tire on a hot day. As the day goes on and it cools off the pressure decreases. The pressure in discs of the back decrease after 2 hours of movement upon waking. This is why exercising early in the morning and or lifting from the floor is not the best time. [11]

One of the key concepts of this program is that of changing motor patterns, the ability to hold the spine in neutral, activate supporting muscles and move in functional patterns of everyday life. [2,4] Examples of this would be lifting clothes from washer to dryer, vacuuming, reaching and pulling, sweeping, putting socks on, getting up from a chair, lifting equipment for work.

The other important aspect of movement is that of a "functional range". [2,8,9] The back can move in many different planes, most notably forward, backwards and twisting. Any structure pushed to its "end point" will stress holding structures and cause failure i.e. injury. It is important when exercising and moving that you stay within your "functional range".

ABOUT THE AUTHORS

**Dr. Douglas J. Taber is a 1998** graduate of New York Chiropractic College in Seneca Falls, NY. He holds a Bachelor of Science Degree through the State University of NY at Fredonia. He is a Fellow of the American Academy of Integrative Medicine, a Fellow of the American Academy of Chiropractic Physicians, a Fellow of the American College of Spine Physicians, a Fellow of the American Board of Disability Analysts, and a Fellow of the American Back Society. He is Board Certified in Pain Management, Motor Vehicle Accident Trauma, and Disability Trauma through the American Academy of Experts in Traumatic Stress, and has achieved Diplomate status with the American Academy of Pain Management, the College of Physicians, and the American Academy of Spine Physicians. He is a member of the American College of Forensic Examiners Institute, a Certified Ergonomics Technician, and is nationally certified as an Independent Chiropractic Examiner by the American Board of Independent Examiners. The doctor maintains a full time private practice as Director of Conservative Care at Comprehensive Spine Care and Orthopedic Surgery in Binghamton, NY. Dr. Taber's articles have been featured in such magazines as Chiropractic Wellness & Fitness, Health & Wellness Magazine, The American Chiropractor and Foot Levelers' "Secrets of My Success" series. He has been featured on various local news programs, newspapers, and the journal Your Health Monthly. His first book, The Back Pain Solution: Unlocking the Spinal Code, was released in May of 2006 and is available worldwide. Spinal Decompression & Stabilization Protocol, Dr. Taber's second book, was released in 2009 and is available at www.decompressionrehab.com and amazon.com. Dr. Taber is a Faculty Member of the New York Chiropractic College Department of Postgraduate and Continuing Education. He regularly lectures to health care providers and the general public on the evidence-based treatment of spine-related disorders and pain syndromes.

 **Dr. Douglas Van Vorst** completed his undergraduate education at Springfield College with a Bachelors degree (B.S.) in biology and graduate education with a doctorate (D.C.) in chiropractic from New York Chiropractic College. He completed his Internship in Long Island at the Levittown Outpatient Facility in Levittown, NY. Dr. Van Vorst has advanced training in Sports Chiropractic having completed a 100 hour Certification Program and passing National Boards as a Certified Chiropractic Sports Practitioner in a post-graduate program with New York Chiropractic College, completing a Hospital Protocols Course and Boards, Certified in Manipulation Under Anesthesia, and completed a 300 hour Diplomate in Chiropractic Rehabilitation. He was the Director of Chiropractic Services at Amsterdam Memorial Hospital in the Division of Chiropractic, Department of Physical Medicine and Rehabilitation having practiced in the Department of Pain Management for the past 9 years, treating complicated neck and low back cases, and now the sole Staff Chiropractor at St. Marys Hospital, Amsterdam N.Y. He is a post graduate instructor sponsored by New York Chiropractic College teaching Lumbar Spine Mechanics and Rehabilitation and an instructor with New York Chiropractic Associations Hospital Protocols Course. Dr. Van Vorst has assisted Dr. Craig Liebenson, world renowned Chiropractic Rehabilitation Specialist and author of **Rehabilitation of the Spine, "A Practitioners Manual"**. Dr. Van Vorst is the Past President of District 10 of the New York Chiropractic Association, an avid sports enthusiast, fisherman, and loving husband and father to his lovely wife Mara and children Emma, Jacob, and Ethan.

PHASE ONE

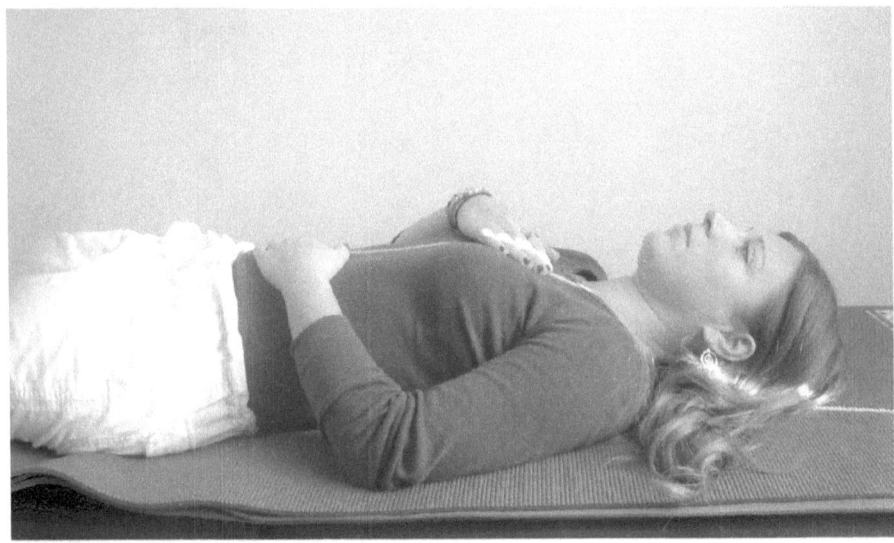

Proper Breathing (belly breathing):

One hand on the chest and the other hand on belly. Take a deep breath and the stomach should rise with no movement from the hand on the chest. This should first be practiced lying on your back and then in a standing position. You should breathe normally when exercising, never holding your breath.

Abdominal Bracing:

Purpose: To activate core and spinal muscles of stability to protect the spine and discs.

Exercise: Lying on your back take a deep breath and bring your belly button in a downward direction. Slowly, isometrically tighten your stomach muscles as you would if someone were to punch you in the stomach. Concentrate and not only tighten the muscles in the middle of the stomach but also on the sides trying to push outwards with these muscles. With the stomach tightened, belly breathe normally.

Notes:

_____

_____

_____

_____

_____

_____

_____

_____

_____

_____

_____

_____

_____

_____

_____

_____

_____

_____

_____

_____

_____

_____

_____

_____

_____

_____

_____

_____

_____

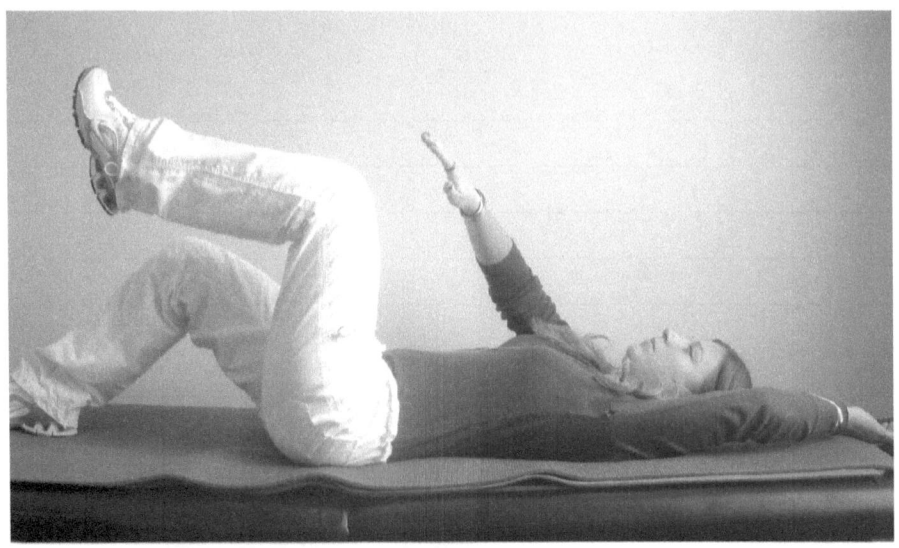

Abdominal Bracing with Dead Bug:

Purpose: To challenge the brace and breathing motor pattern.

Exercise: Beginner: In a supine position (on your back), knees up, arms overhead, brace and breathe, slowly lifting each arm overhead one at a time and slowly lower making sure the abdominal brace is not lost and not holding your breath.

Notes:

_____
_____
_____
_____
_____
_____
_____
_____
_____
_____
_____
_____
_____
_____
_____
_____
_____
_____
_____
_____
_____
_____
_____
_____
_____
_____
_____

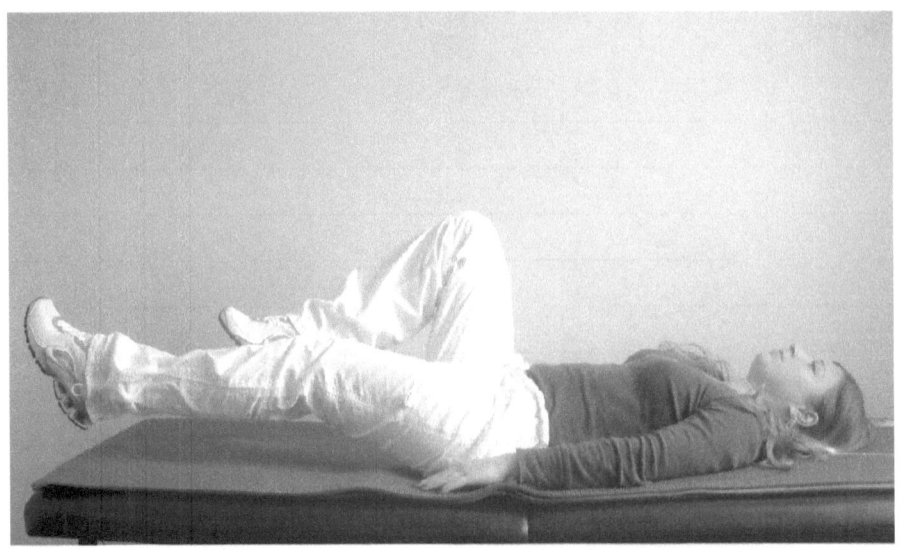

Abdominal Bracing with Dead Bug and Leg Extension:

Purpose: To challenge the brace and breathing motor pattern.

Exercise: Beginner: Laying on your back, bracing and breathing, lift one foot off the table about 2 inches and fully extend the leg, bring leg back and extend the other leg.

Notes:

_____

_____

_____

_____

_____

_____

_____

_____

_____

_____

_____

_____

_____

_____

_____

_____

_____

_____

_____

_____

_____

_____

_____

_____

_____

_____

PHASE TWO

Bird Dog, Neutral Spine with Abdominal Brace:

On your hands and knees, head in neutral position, hold lumbar lordosis and activate your abdominal muscles in a brace. Belly-breathe normally. Extend one arm out and then the other arm, one at a time, holding your pelvis level and not losing your lordosis and brace. Now try extending one leg at a time. The arms and legs should reach out level with the body with no torque of the body or pelvic area. Once this is accomplished now try the opposite arm and leg.

Notes:

_____
_____
_____
_____
_____
_____
_____
_____
_____
_____
_____
_____
_____
_____
_____
_____
_____
_____
_____
_____
_____
_____
_____
_____
_____
_____
_____
_____

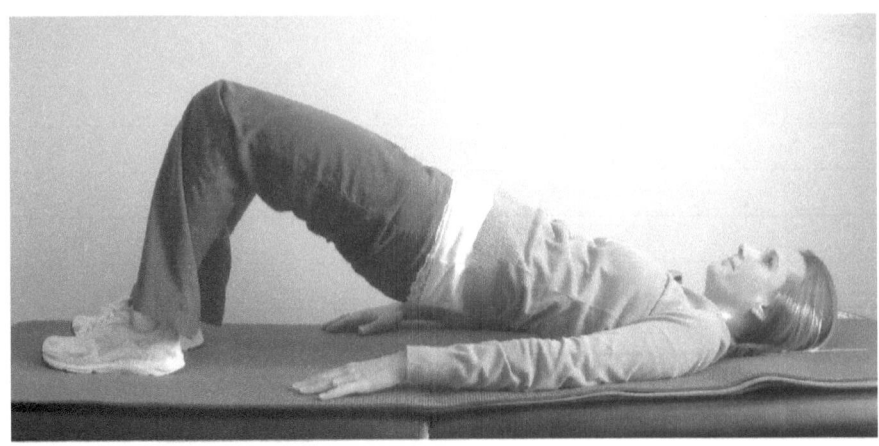

Bridge with Abdominal Brace:

Lying on back with knees up and feet down, hold abdominal brace, squeeze buttock and lift buttock up holding lumbar lordosis. Concentrate and contract quads and not muscles.

The next progression for the bridge exercise would be to lift each foot 2 inches off the floor or table, one at a time, holding the abdominal brace with proper breathing. Once this is accomplished, extend each leg outward, one at a time. Remember all exercises should be done slowly and held steady, with no shifting of the body left or right.

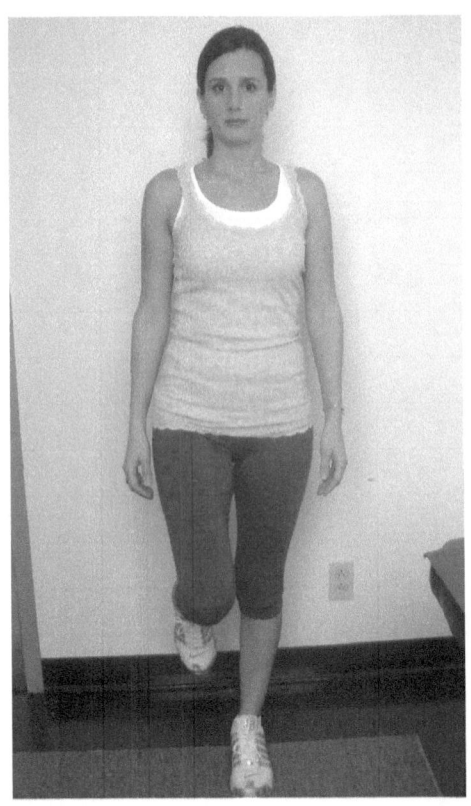

One Leg Stands:

Purpose: Challenge balance, activating propioception of the feet. Kinetic chain stability starts in the feet. In a standing position, stand on one foot lifting the other leg to 90 degrees. Grip the ground on the standing foot with toes and outside of foot. This should be done with each foot and held for approximately 29 seconds.

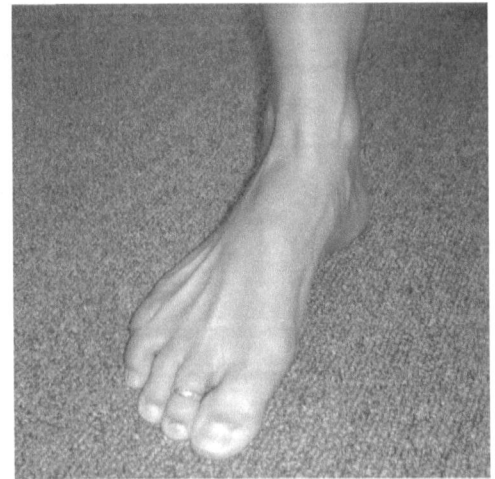

Note: Your doctor or therapist should have shown you a "short foot" exercise.

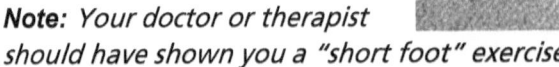

Notes:

_____
_____
_____
_____
_____
_____
_____
_____
_____
_____
_____
_____
_____
_____
_____
_____
_____
_____
_____
_____
_____
_____
_____
_____
_____
_____
_____

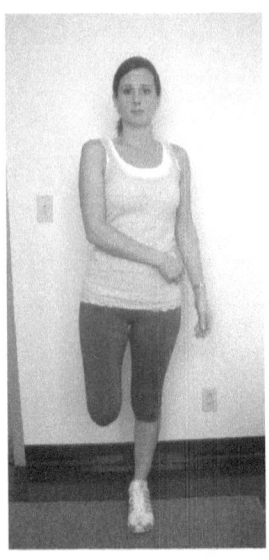 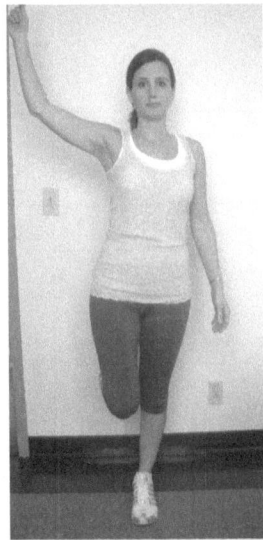 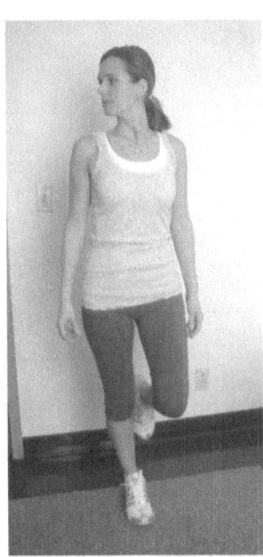

One Leg Stands with Challenges:

Purpose: Further challenge balance and stability and activate proprioception of the feet.

Exercise: The next progression would be to stand on one foot. Take your right arm and reach for your opposite front pocket. Pretend you are pulling a sword out of your pocket and raising the sword above your head. Do this very slowly, challenging your weight distribution from one side of your body to the other.

The final challenge would be to slowly turn your head left and then right. These exercises should be done on one foot and then the other. Do not proceed with the next progression until you have mastered the exercise at hand.

Notes:

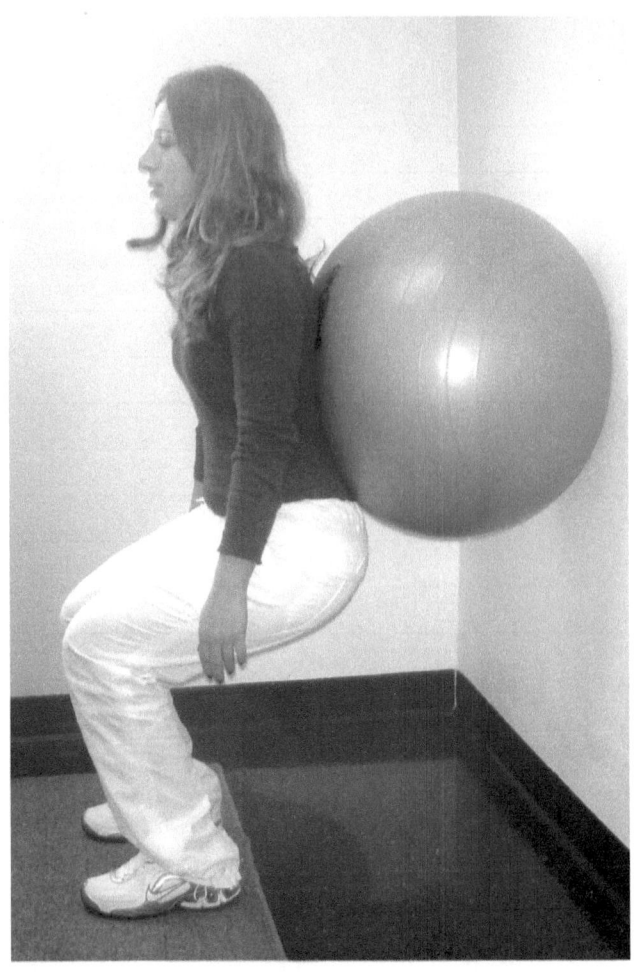

Squat with PhysioBall:

Using the physioball, put the ball against the wall with your back towards the ball. The ball should be in the lordosis of your back. Holding abdominal brace and good belly breathing, squat to 90 degrees tucking your buttock under the ball.

Notes:

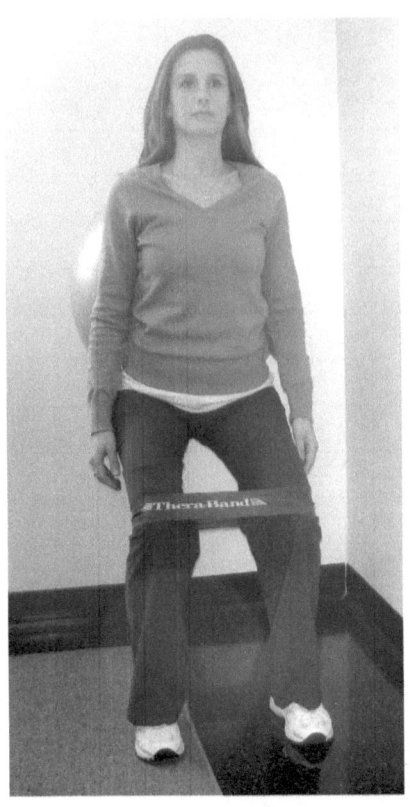

Squat with PhysioBall and Knee Tubing:

Purpose: In the position shown in the above photo, with stabilization of knees and activation of glute medius.

Exercise: If you have a knee problem or your doctor feels you need activation of the glute medius (buttock muscle) you may need to use tubing around your knees to stabilize the knees or activate the glute medius.

Notes:

_____

_____

_____

_____

_____

_____

_____

_____

_____

_____

_____

_____

_____

_____

_____

_____

_____

_____

_____

_____

_____

_____

_____

_____

_____

_____

_____

_____

PHASE THREE

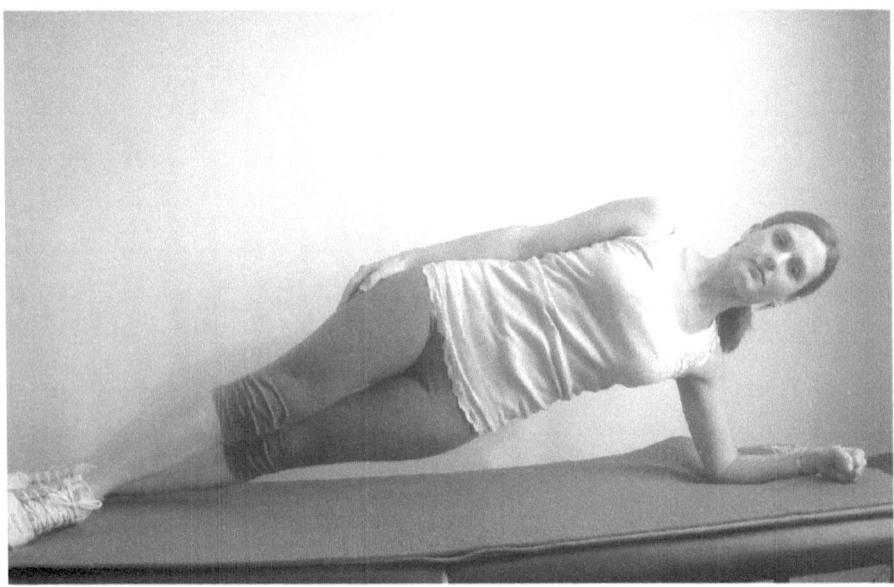

Side Bridges:

Purpose: To activate and strengthen the lateral stabilizers of the core/spine (Quadratus Lumborum) with minimal disc loading.

Beginner Exercise: In a side lying position with knees bent and body in a straight line, lean on the elbow with shoulder in a "locked in" position, directly under shoulder. Lift your hips and move your pelvis forward with a hip hinge movement. Remember to activate the abdominal muscles and breathe normally. The upright position should be held for a count of 2 and slowly return the hip to the floor. This should be done on both the left and right sides.

Notes:

_____

_____

_____

_____

_____

_____

_____

_____

_____

_____

_____

_____

_____

_____

_____

_____

_____

_____

_____

_____

_____

_____

_____

_____

_____

_____

_____

_____

_____

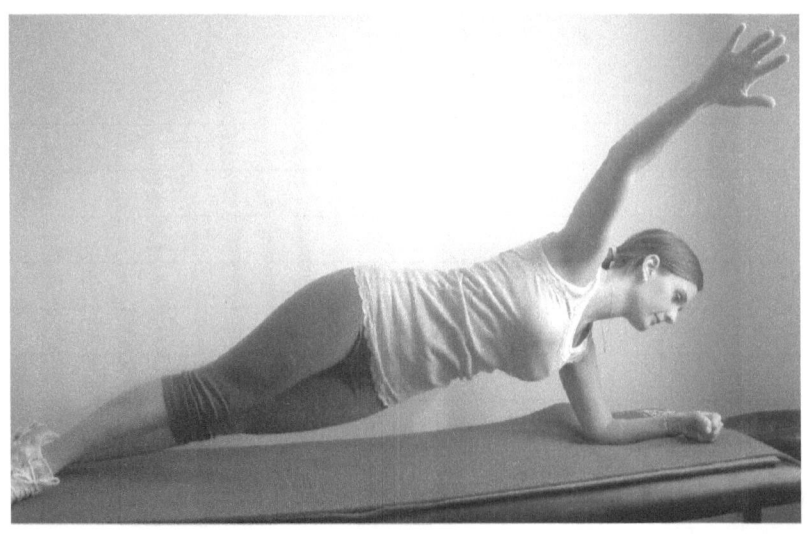

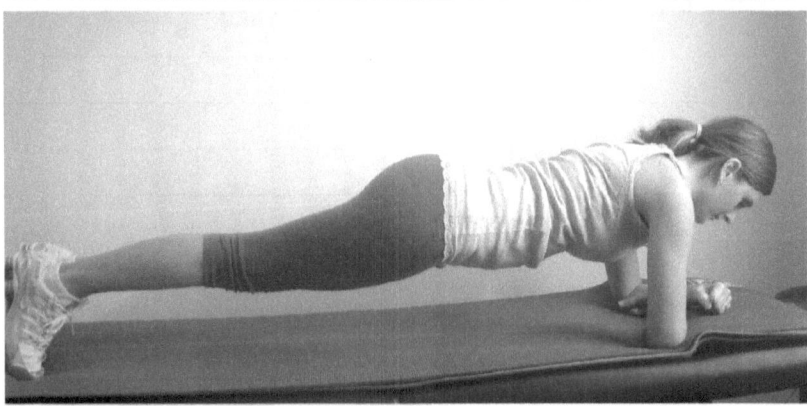

Side Bridges to Plank Rolls:

Advanced Exercise: In a side lying position, instead of knees bent, keep the legs straight. You will now be lifting the entire body in a plank position, weight bearing on the elbow and the foot/ankle area. This should be held for 2 seconds and slowly lowered.

Plank: Hold the upright side bridge position. With your free arm, reach towards the ceiling. Slowly rotate slightly backwards and then slightly forwards, making sure to keep the body super stiff and breathe normally. Once you are able to do this, slowly rotate your body forward onto both elbows and hold the plank position, rolling to the other elbow and reaching upward with the top arm.

Notes:

_____

_____

_____

_____

_____

_____

_____

_____

_____

_____

_____

_____

_____

_____

_____

_____

_____

_____

_____

_____

_____

_____

_____

_____

_____

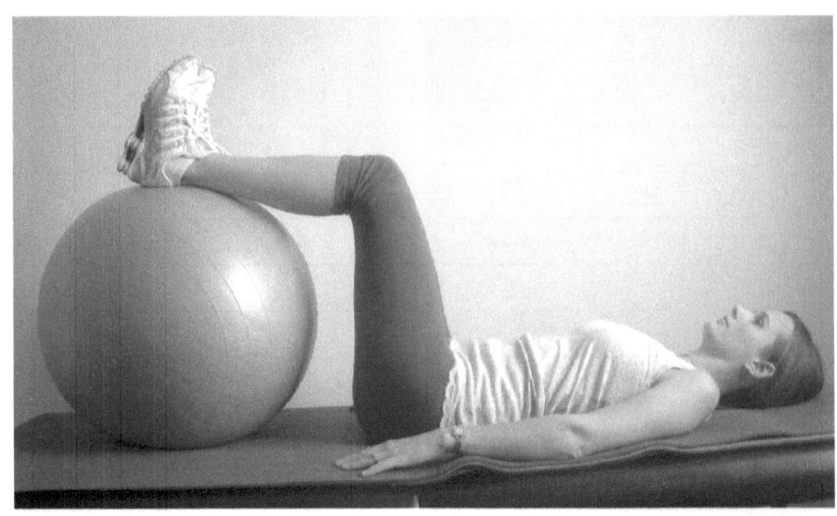

Bridge with Feet on PhysioBall:

Purpose: Core stability, quad activation.

Exercise: Lying on your back put your feet/heels on the ball in a 90/90 position. Make sure your feet are approximately 4-5 inches a part. Brace, tighten the buttock muscles and lift your buttock off the ground. Concentrate on digging your heels into the ball and bringing your toes towards your head. Again, feel the contraction of your quads and not the hamstring muscles. Hold your body rigid.

Notes:

_____

_____

_____

_____

_____

_____

_____

_____

_____

_____

_____

_____

_____

_____

_____

_____

_____

_____

_____

_____

_____

_____

_____

_____

_____

_____

_____

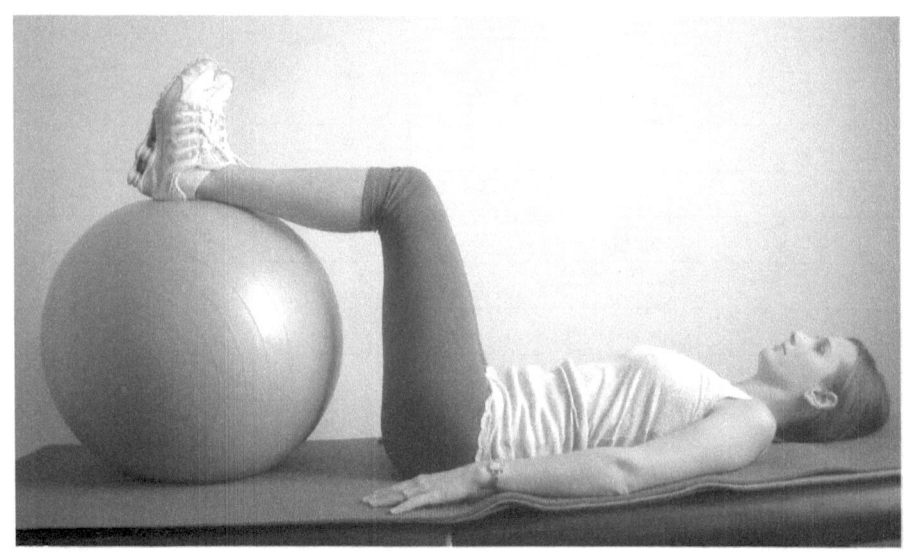

Hamstring Curls with PhysioBall :

Purpose: Core stability and hamstring strengtening in the position shown in the above photo, but extending the legs straight and then curling the ball back towards your buttock. This exercise is for the hamstrings. Concentrate on holding the body stiff and steady with no sway left or right.

Notes:

_____

_____

_____

_____

_____

_____

_____

_____

_____

_____

_____

_____

_____

_____

_____

_____

_____

_____

_____

_____

_____

_____

_____

_____

_____

_____

_____

_____

_____

_____

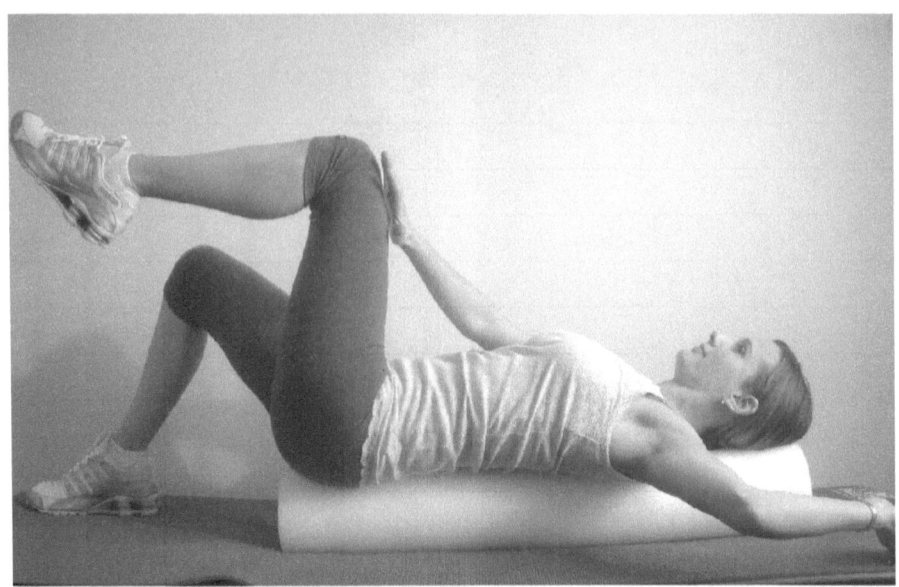

Dead Bug with Foam Roller:

Purpose: Core activation.

Exercise: Using a ½ foam roller lie on your back so the roller supports your head and back. Your feet and hands will contact the floor. While bracing the abdomen and belly breathing, slowly lift each foot independently holding the body rigid. Count slowly, one, two and then switch feet. This exercise can progress to a full dead bug exercise touching your hand to the opposing knee and then performing the above progression using a full foam roller.

Notes:

PhysioBall Curl Ups / Modified McGill:

Purpose: To strengthen and activate the core muscles with minimal disc pressure.

Exercise: Lying on the physioball with your knees at 90 degrees and hands supporting the neck, keep your head in a neutral position and slowly lift your chest upwards approximately 2 inches. Do not curl the neck up or the spine. This exercise should be done very slowly working both the up motion and slowly going down. Remember it is the quality of the exercise not the number of curl ups you can do.

Notes:

_____

_____

_____

_____

_____

_____

_____

_____

_____

_____

_____

_____

_____

_____

_____

_____

_____

_____

_____

_____

_____

_____

_____

_____

_____

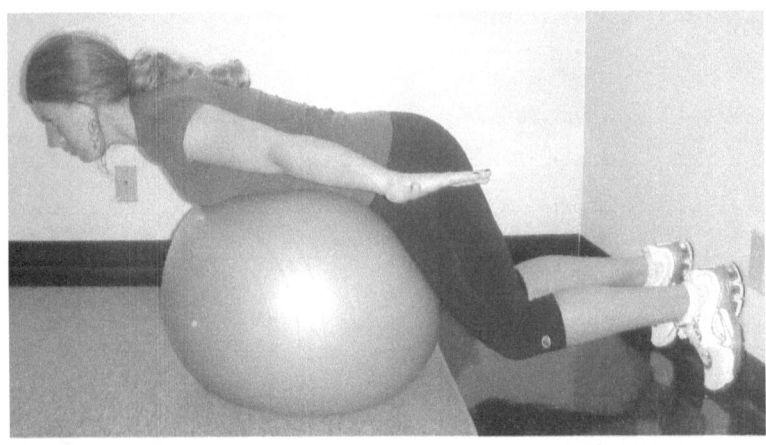

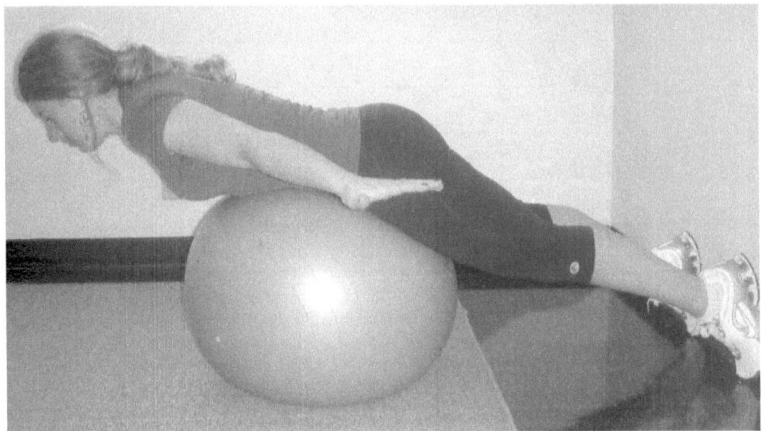

Superman / Superwoman:

Purpose: This is an extensor activating exercise to work the muscles on the back of the body the backside of the guide wires protecting the spine.

Exercise: Lying on top of the ball, feet against the wall and spread approximately 1-2 ft a part, hold the head in neutral, arms behind you with palms down, knees slightly off the ground. This is your start position. Slowly extend your legs outward, pushing off the wall with your toes and spreading your fingers. This is an exercise for the extensors which are your calves, buttock, back muscles and triceps in your arms. Slowly push away from the wall and return to your start position. Do not lift your head or arch your back. Think of it as a squat against the wall.

Notes:

_____

_____

_____

_____

_____

_____

_____

_____

_____

_____

_____

_____

_____

_____

_____

_____

_____

_____

_____

_____

_____

_____

_____

_____

_____

_____

_____

_____

Designing a Star Pattern:

Put duct tape on the ground in a star pattern, two pieces 4ft long in a criss cross at 90 degrees and the other two pieces obliquely, forming a star. If possible put the star pattern on the floor near a door so the same pattern can be used for the exercises below.

Lunges:

Purpose: Dynamic, functional stability, utilizing proper breathing, hip hinge and squat.

Exercise: In a standing position, with one foot, take a step forward. Make sure both knees are bent on both the forward stepping leg and the back leg. Look down and make sure the forward legs knee is not beyond the toe. Stay balanced between your two feet. Don't lunge in a straight line. Lunge slightly left with the left foot and right with the right foot. Again, lunge with abdominals braced, holding the low back in lordosis. Try now lunging in a star pattern with the right foot lunging to each piece of tape on the right and left foot to the left, pivoting on the back foot.

Notes:

_____

_____

_____

_____

_____

_____

_____

_____

_____

_____

_____

_____

_____

_____

_____

_____

_____

_____

_____

_____

_____

_____

_____

_____

_____

Lunges with Medicine Ball Reach

Purpose: To challenge the lunge and center of balance, further activate the core and to mimic a movement of properly moving an item with weight from one place to the other in a multiplaner fashion.

Exercise: In the position shown in the above photo, but using light 3-4 lbs dumbbells or a medicine ball, hold the weight in to your chest and as you lunge reach with the dumbbells or medicine ball. Remember to brace as you do this, hold lordosis and keep the body balanced, not going outside your center of gravity.

Notes:

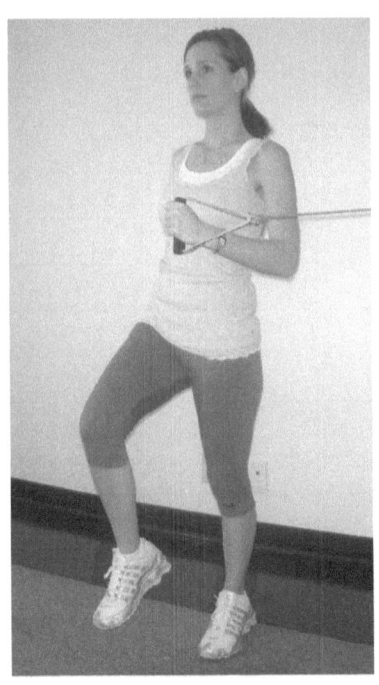

Tubing Star Exercises with/without Punch:

Purpose: Challenge trunk stability while rotating, mimicking an activity like golf, baseball, sweeping.

Exercise: Use the pattern shown in the above photo. This is a dynamic, functional lunge exercise which mimics the movement of a baseball pitcher. The key to this exercise is to have the upperbody move in conjunction with the lower body. Put tubing end in door or attach to immovable object. Stand in the Star pattern with toes on a line, the start position. Hold the handle by your side with slack out of tubing, at belly button height. Put all your weight on your back foot, rotating on the back toe and lunging forward now perpendicular to the line your toes were on. You should still have the tubing in the area of your belly button and end in a lunge position as above. Again, do not stand in a straight line, have a steady base. Slowly come back to the start position. Once stable with the hand held at belly button height, perform a punch forward as you lunge. Slowly return to start position. This can be done both with the right hand and turning around to the other side of the line, with the left hand.

Notes:

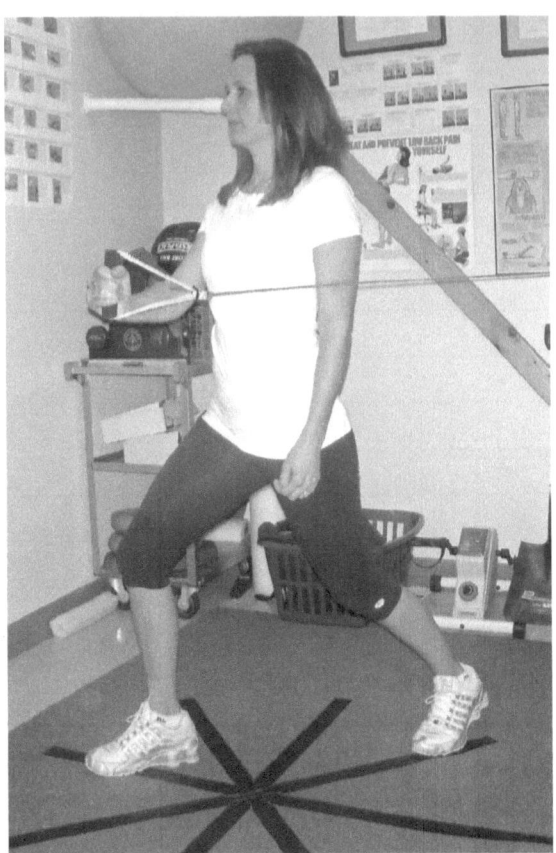

Star Backhand Punch:

Purpose: Working the muscles and balance on the "backside".

Exercise: In start position with the right foot as the back foot, now hold the handle in the left hand. Lunge as shown in the above photo, but this time perform a backhand or Frisbee type motion with your left hand, again at belly button height. This will work the opposing muscles on the backside.

TIPS

Rules for Lifting

1. Assess the load.

2. Prepare yourself.

3. Stand with your feet shoulder width apart.

4. Bend with the knees, keeping the natural curves of your back in alignment.

5. Grip the object, pulling it close to your body.

6. Tighten your stomach muscles.

7. Keeping your back in alignment, lift with the legs in a slow, smooth motion.

8. While carrying a load, keep it close to your body, and do not twist or bend at the waist.

9. Put the object down the same way in which you picked it up.

Practical Pearls

- Keep items, especially heavy items, up off the floor.

- Tighten your abdominal muscles every time you initiate a movement.

- Chin up buttock out when bending.

- Wait 2 hours prior to any exercising or lifting that involves bending to the floor.

- Stay upright keeping the curve in your back, put your foot on a chair, when putting your socks and shoes on.

- Don't do toe touches.

- Avoid dead lifts, squats and rowing machines if you have a history of back problems.

CONCLUSION

Training Schedule

_____

_____

_____

_____

_____

_____

_____

_____

_____

_____

_____

_____

_____

_____

_____

_____

_____

_____

_____

_____

_____

_____

_____

_____

_____

_____

_____

BIBLIOGRAPHY

1. Panjabi MM, *The Stabilizing system of the spine. Part 1: Function, dysfunction, adaptation, and enhancement.* J Spinal Disord 1992; 5: 383-389.

2. McGill S., *Low Back Disorders, Evidence based prevention and rehabilitation.* Liberty of Congress Publication, 1997.

3. Altered trunk muscle recruitment in people with low back pain with upper limb movement at different speeds*1, *2. *Archives of Physical Medicine and Rehabilitation,* Volume 80, Issue 9, Pages 1005-1012. P. Hodges, C. Richardson.

4. *Inefficient muscular stabilization of the lumbar spine associated with low back pain: a motor control evaluation of transverse abdominis,* Hodges, Paul W., Richardson, Carolyn A. Spine, Nov 1996, Vol 21-issue 22, pp 2640-2650.

5. Does strengthening the abdominal muscles prevent low back pain, a Randomized Control Trial, A, Helewa, et al. *Journal of Rheumatology* 1999; 26:1808-15.

6. Deutsch, FE. Isolated lumbar strengthening in the rehabilitation of chronic low back pain. *Journal of Manipulative Physiological Therapy.* 1996 Feb; 19(2): 124-33.

7. Cholewicki J, McGill SM, Mechanical stability of the in vivo lumbar spine: implications for injury and chronic low back pain. *Clinical Biomechanics.* 1996; II:I-15.

8. Liebenson, C., *Functional Stability Training in Rehabilitation of the Spine; Practitioners Manual,* Lippincott Williams and Wilkins, 2nd edition, Philadelphia 2007, pgs 612-660

9. Richardson C, Jull G, Hodges P, Hides J., *Therapeutic exercise for spinal segmental stabilization in low back pain.* Edinburgh: Churchill-Livingston, 1999 pgs 151-164

10. Kahl M., Herring S., *Rehabilitation of Lumbar Spine Injuries in Functional Rehabilitation of Sports and Musculoskeletal Injuries, Aspen Pub. Maryland,* 1998 pgs.188-195

11. Snook SH, Webster BS, McGorry RW, *The reduction of chronic, nonspecific low back pain through the control of early morning lumbar flexion: 3-year follow-up.* Department of Environmental Health, Harvard School of Public Health, Boston, Massachusetts, USA. stover@att.net

12. Elphinston, J., *Stability, Sport and Performance Movement,* Lotus Publishing, Chichester, England 2008

www.ingramcontent.com/pod-product-compliance
Lightning Source LLC
Chambersburg PA
CBHW021043180526
45163CB00005B/2261